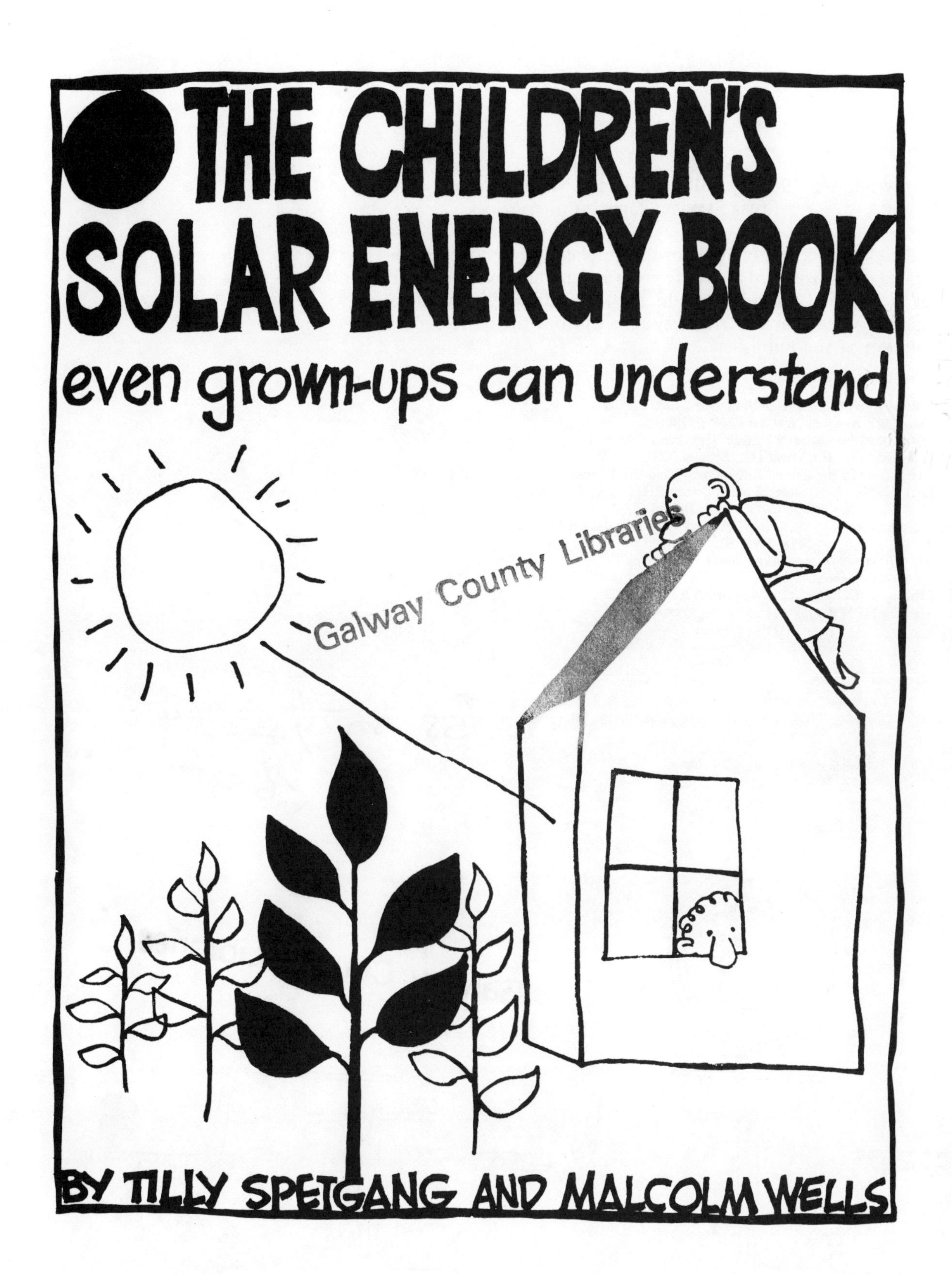

Sterling Publishing Co., Inc. New York

Published by Sterling Publishing Co., Inc.
Two Park Avenue, New York, N.Y. 10016
Distributed in Australia by Oak Tree Press Co., Ltd.
P.O. Box J34, Brickfield Hill, Sydney 2000, N.S.W.
Distributed in the United Kingdom by Blandford Press
Link House, West Street, Poole, Dorset BH15 1LL, England
Distributed in Canada by Oak Tree Press Ltd.
c/o Canadian Manda Group, 215 Lakeshore Boulevard East
Toronto, Ontario M5A 3W9
Manufactured in the United States of America

Library of Congress Catalog Card No.: 81-85022
Sterling ISBN 0-8069-3118-3 Trade
3119-1 Library
7584-9 Paper

CONTENTS

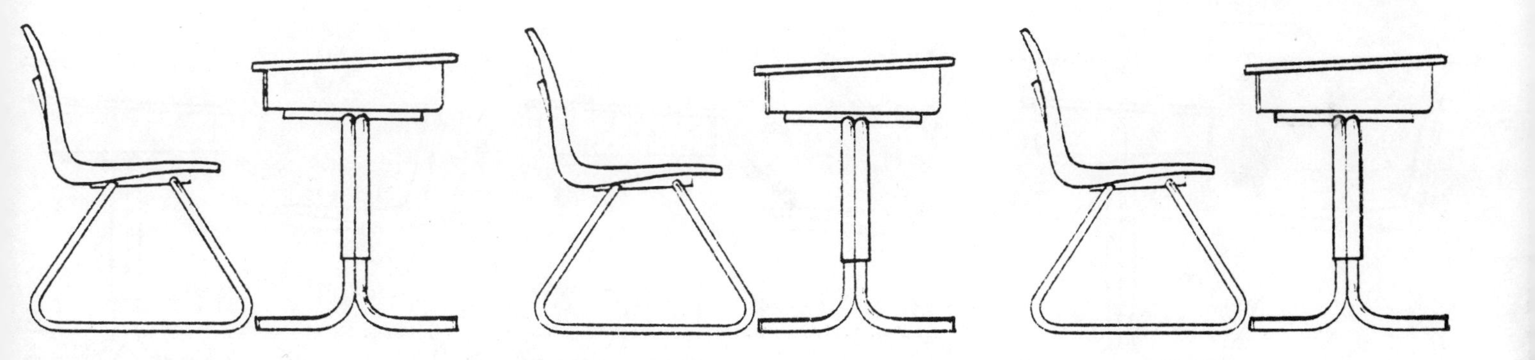

Galway County Libraries

WILLIAM SHURCLIFF AS A BOY.

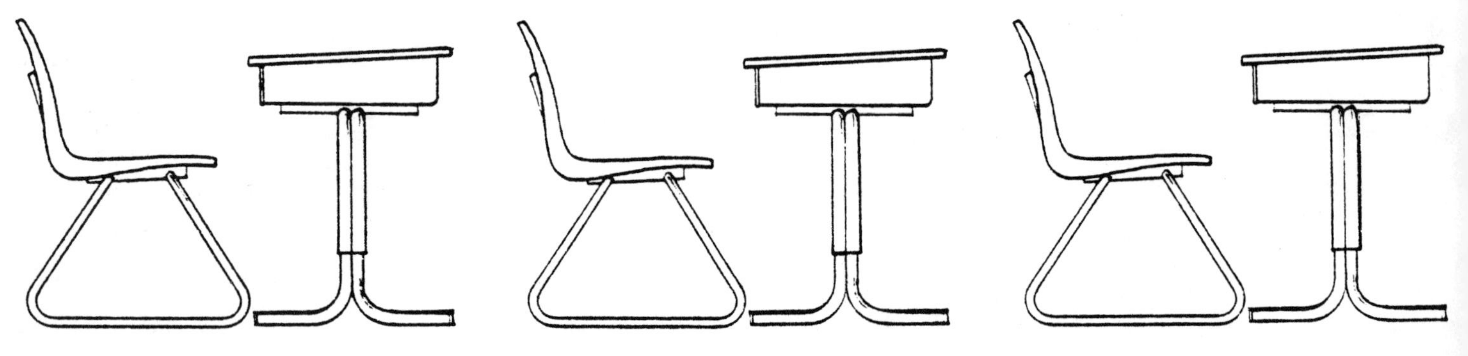

INJUSTICE, HUNGER AND YOUR GLORIOUS FUTURE

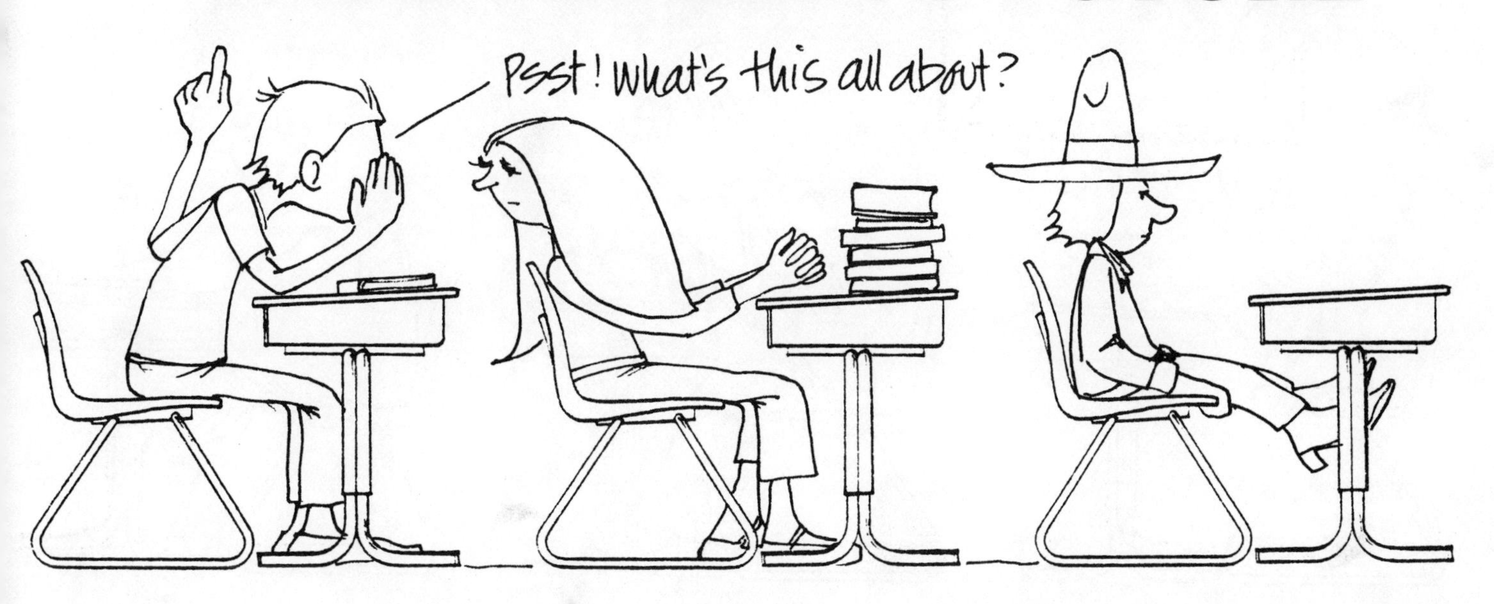

This world is yours.

That's very generous of you.
Thanks a lot, Mrs. Robinson.

Oh, no!

This beautiful blue-green planet of mountains and oceans and skyscrapers and great beasts of the jungles. . . . all yours.

Have you ever thought about it that way?

You may be about 12 years old and not feel as if you count in any special way outside your family.

But, in the wink of an eye, you will be 18, 23, 32.

Then you will be
in the position of electing the politicians who make the rules of the land.

You may own property, pay taxes, and help decide how that tax money should be spent.

You will have power!

As you grow older, you will begin to realize how many cruel and unjust things go on in life.

You probably will
try to change some of these injustices.

It may be that
people are hungry and are not being fed. You may fight to stop the killing of giant whales, the clubbing of baby seals.

You may come to feel that the spread of nuclear power means deadly destruction.

As an adult you will be able to give your time, your creative efforts, and your money to help solve some of the world's problems.

One of the most serious of those problems that YOU will have to deal with will be . . . . energy.

You've heard about the energy crunch.

Your mother lowers the temperature in the house. . .
to save energy.

Your father car-pools with two other people. . . to save energy.

You are asked to turn off lights and the television set when you leave a room. . . to save energy.

Why do we HAVE to save energy?

Why can't we simply
fill the gas tank in the car, the oil tank in the basement, or move a switch so we have electricity whenever we want it? What's all the noise about, anyway?

Well, to be truthful, most people want to save energy because it costs so much.

I always thought that it was its discovery, extraction, processing, distribution, and conversion rather than the energy itself that was so expensive.

But the major underlying problem is oil.

We're running out of it.

Good.
That'll get rid
of the oil problem.

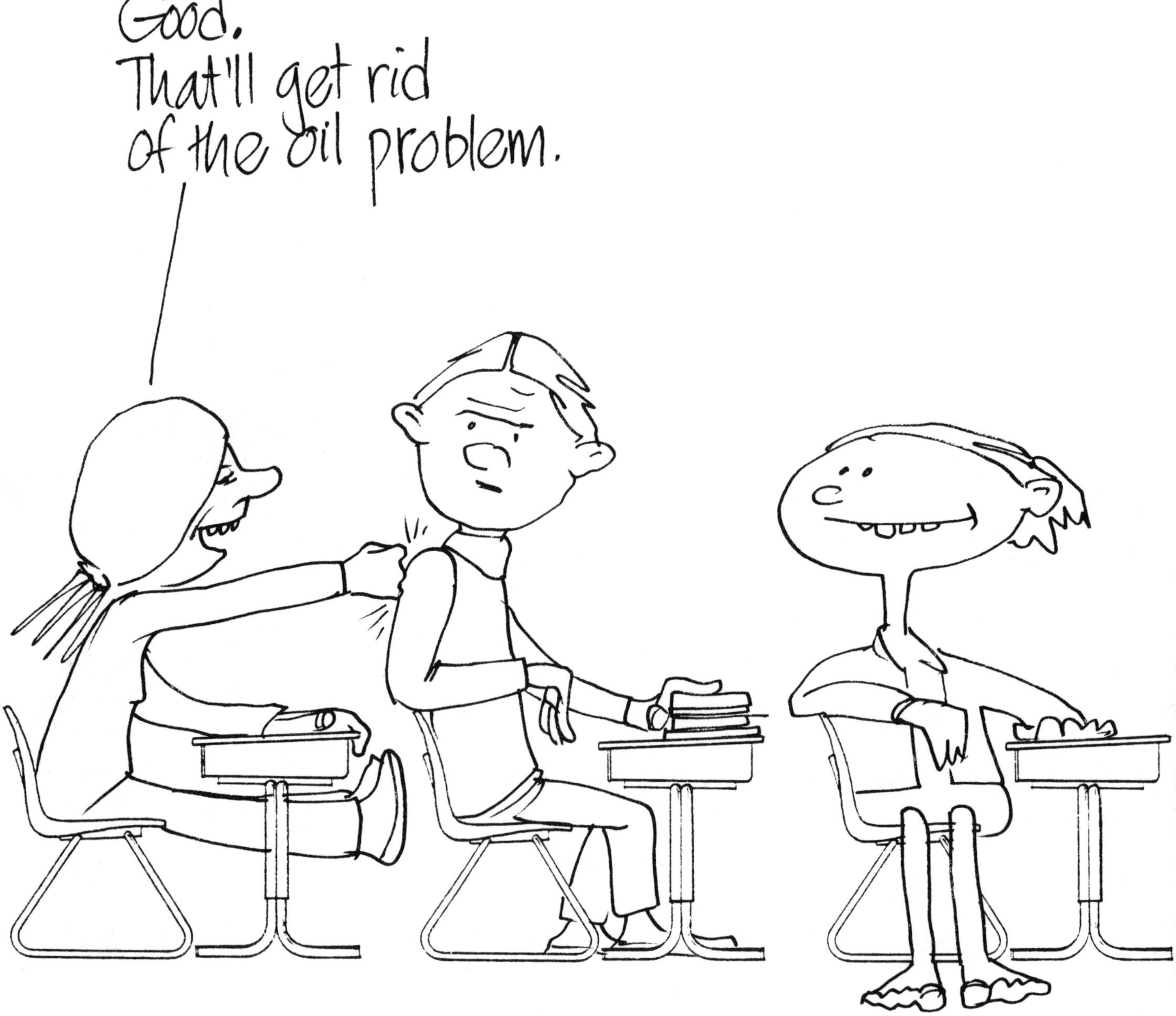

It takes an enormously long time for oil to be created in the earth, where heat and pressure help to form it.

The oil
we've pumped out in the past 10 years may have taken millions of years to form.

You can plainly see that we cannot keep using oil this recklessly or it will be gone in the future, when you, and your children, will need it.

Oil is very important.

It is used in making medicine, for instance.

It keeps giant machinery running.

I'll sleep a lot better, knowing that.
RICHARD

In the form of
gas it makes cars move; it is important to farming because fertilizer is made from it, and it helps make electricity.

All over the world right now, methods are being studied of creating energy without depending on oil.

One of these
ways is solar energy. . . using the sun for heat.

How *what* all began?

HOW IT ALL BEGAN

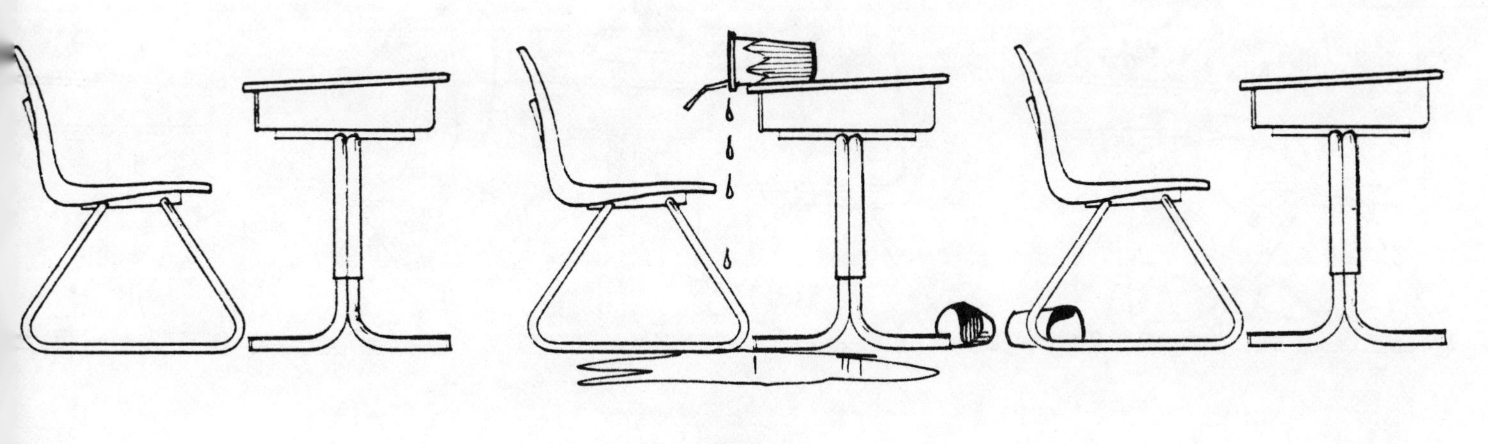

If you stopped a caveman and asked him (assuming there was a language you both spoke) where he learned about passive solar heating, he wouldn't know what you were talking about.

But the caveman used the sun in a smart manner. He moved into caves facing the south, from where the sun shines.

This gave him sunlight which baked the floor and walls of his stone cave during a good part of the day. Then, at nightfall, when the weather turned cold, those walls and floor gave back much of the same heat, warming the caveman and his family.

In America's very early days, the Pueblo Indians built their adobe rooms into the south face of cliffs. They used the sun exactly the way cave dwellers did. . . for stored heat in walls and floor.

How did they know about it?

Simple. They must have
felt with their hands how stone sucked up heat during the day, then released it at night.

No. I think history tells us they always felt stone with their elbows.
V

They were smart people, and
they applied what they knew to their dwellings.

In a major exposition (like a world's fair) in France during the late 1890's, one of its most exciting exhibits showed how a piece of machinery could be run by the sun.

A large, shiny metal parabola (nothing more than a large, curved dish) focused the sun's rays on a black tank of water. When the water in the tank boiled, it gave off steam.

The steam was sent through a pipe to run a small steam engine which, in turn, was the driving power for a printing press.

The press printed leaflets...

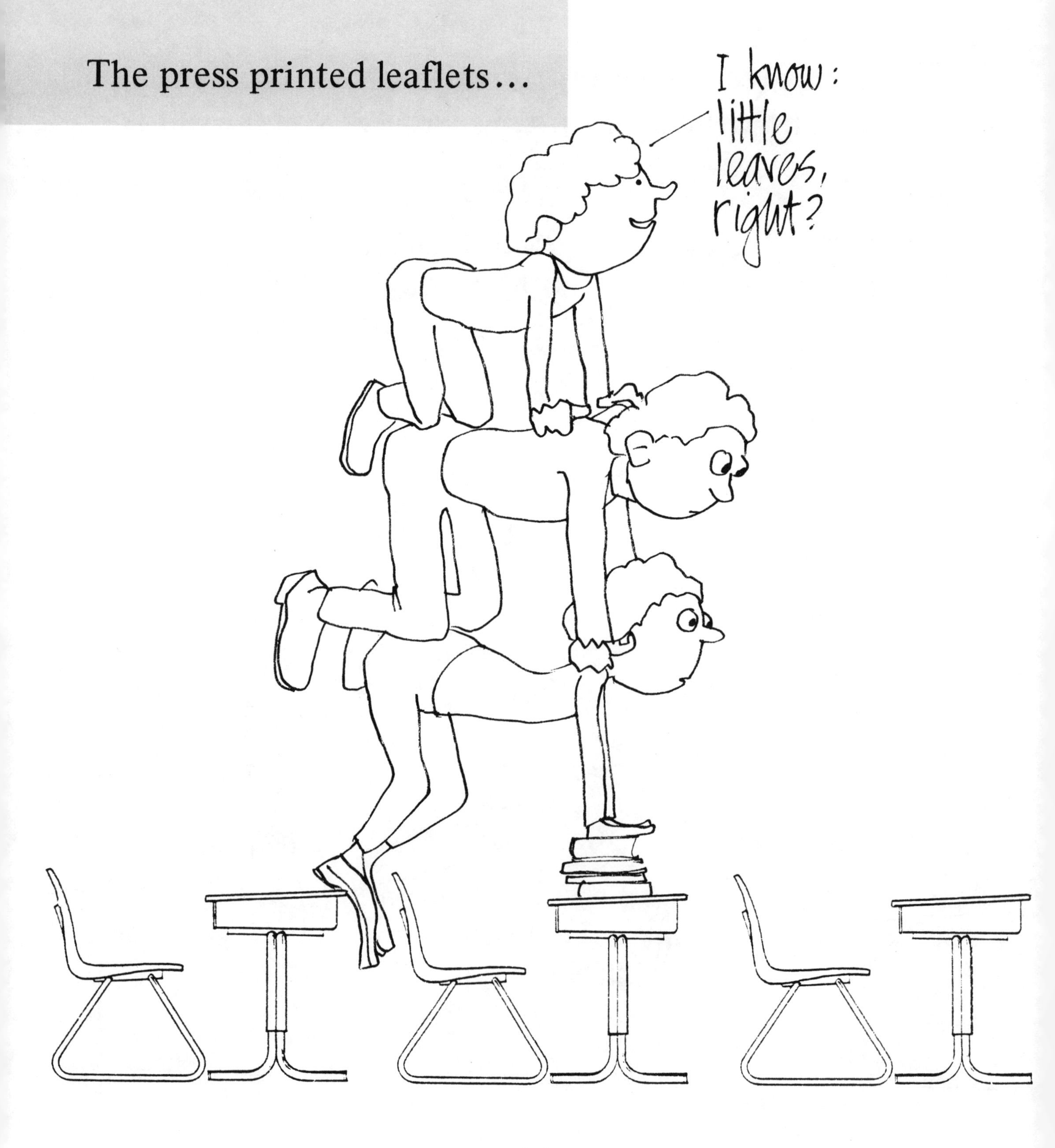

(many viewers took them as souvenirs) which explained that all the work of printing the leaflet had been done by the sun, without electricity or other power sources.

It was then, and still is, a great source of wonder.

Also, in the late 1890's, in the southern parts of the United States, several types of solar systems were sold to people in Florida and California.

These were roof mounted
solar collectors that heated water used for dishwashing, baths, and laundry.

Heating water by using the sun's rays was a great deal easier for people living in those hot southern states than chopping wood, building a fire, and keeping a stove going in the heat of the summer.

Wouldn't it have been easier than chopping wood almost anywhere?

So solar energy was very popular then.

Then the oil problem arrived,
I'll bet.
SAM

But in the early 1900's, oil and gas were discovered (the very fuels that are becoming scarce now).

Men quickly
learned that all they had to do was run gas through a pipe, then light the gas as it came out.

It was cheap. It was easy. Solar energy was put aside as being too bothersome.

In the late 1950's, rumblings were heard about possible shortages in fuels.

Most people pooh-poohed the rumors because the idea was so strange.

Hadn't there always been plenty of oil and gas?

No!
They were discovered only 60 years earlier, remember? You told us yourself.

Wouldn't they continue. . . forever?

But a few men and women around the country listened very carefully.

They were known as conservationists. They began to change their lifestyles so they would use less electricity, water, coal, gas and oil.

And out of their
searching for ways of saving fuels grew a renewed interest in solar energy.

These early conservationists studied, experimented (one woman in Princeton placed home-made solar panels on her garage roof that heated the oven in her house, enabling her to cook and bake by the sun's free heat), and began to put their money and thoughts into solar energy.

Today there are many good manufacturers of solar energy systems, many fine engineers and architects who are able to design solar systems properly.

Industry, as well
as government, is studying solar energy seriously as a way to stop depending on foreign oil.

You will benefit from all this solar activity. And some day, when you are an adult, you may remember learning your solar energy basics.

Enjoy.

ACTIVE SOLAR SYSTEMS

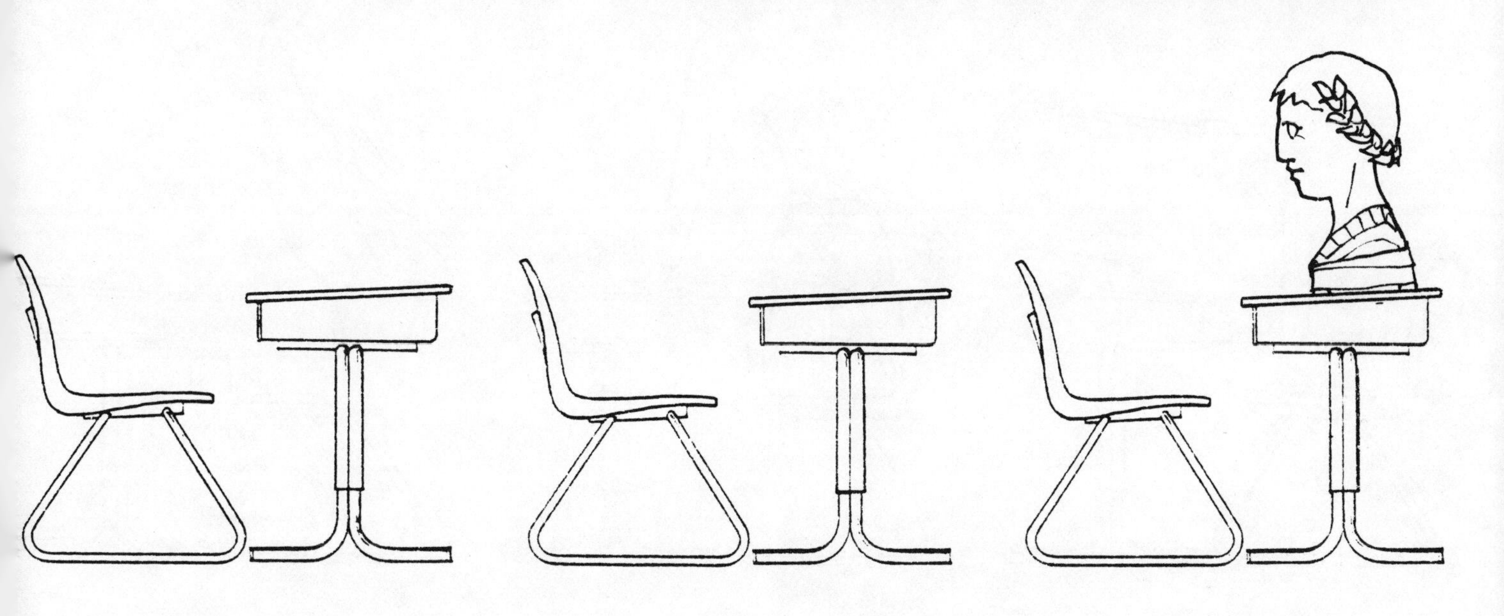

Energy from the sun is free.

But there's a catch, right? You have to do something expensive in order to use it.

It doesn't pollute the air as oil and coal do.

It's a natural way of heating air and water. So you may wonder why EVERYBODY doesn't use solar energy instead of oil.

Well, in some countries, everybody DOES.

In Israel
and Japan, for instance, solar panels dot roofs everywhere.

That's because those countries have no oil or coal of their own and must purchase energy from other countries at very high prices. America is rich in coal, imports less than half the oil it uses, so it doesn't need solar energy as badly as Japan and Israel.

Another reason everybody doesn't use solar energy instead of oil is that it's a fairly new idea.

People don't
quite know how it works and are a bit nervous about anything they don't understand. But in YOUR world. . . the near tomorrow. . . there will be many solar buildings.

Your home will probably use solar energy. So you should know and understand what it is and how it works.

Have you learned that the sun is a star, the same kind of star you see crowding the heavens on a dark night?

It is a light, heat, and energy star composed of burning gases.

The sun is the star nearest to the earth, which turns about it, gathering up heat and light from this flaming neighbor 93,000,000 miles away.

So there is the sun, giving its heat away free. But this is the problem: How do you catch sunbeams to make them work for you?

One way is through the use of a solar collector, which is nothing more than a different way of saying sun collector; It collects the sun's heat.

It does it the way a greenhouse works.

A greenhouse is made mostly of glass to catch hot sunshine through its roof so flowers, plants and vegetables can grow in it all through the bitter winter months.

A solar collector is usually a flat box that lies on the roof of the house.

The box can be made of metal
with a glass cover, just like a little flat greenhouse. But, when the sun comes through the glass and hits the inside of the box, it stops being sunlight and turns into heat.

The inside of the box, which is black, gets very hot. It stays hot until water in special pipes flows over it to cool it off.

Why cool it off if you went to all that trouble to get it hot?

Air can cool it, too.

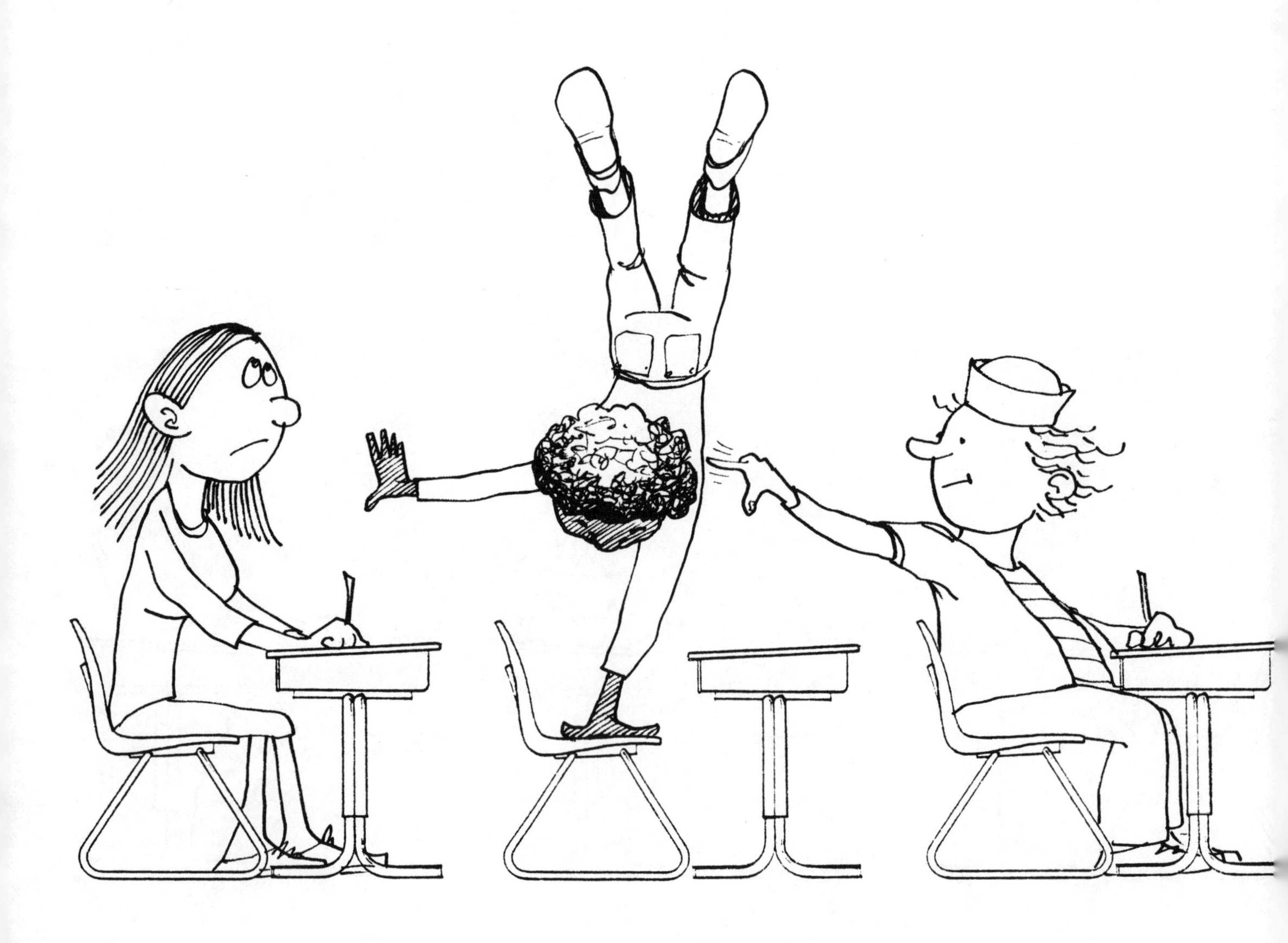
she's not listening to me.

The heat is now in the water (or in the air, if you used air, but let's talk about water).

The hot water flows into a tank where it is stored.

Then, when the house gets cool and heat is needed, a fan blows air past the hot water tank.

This warms the air which then flows all through the house, eventually coming out through the grille in your bedroom.

So there we have it: the collection, transformation, and use of energy. The energy which traveled to earth only a few hours earlier as sunlight is now warming my bod. What happens next? Does the energy simply disappear? Or do I release it as heat through my breath and sweat?

PASSIVE SOLAR SYSTEMS

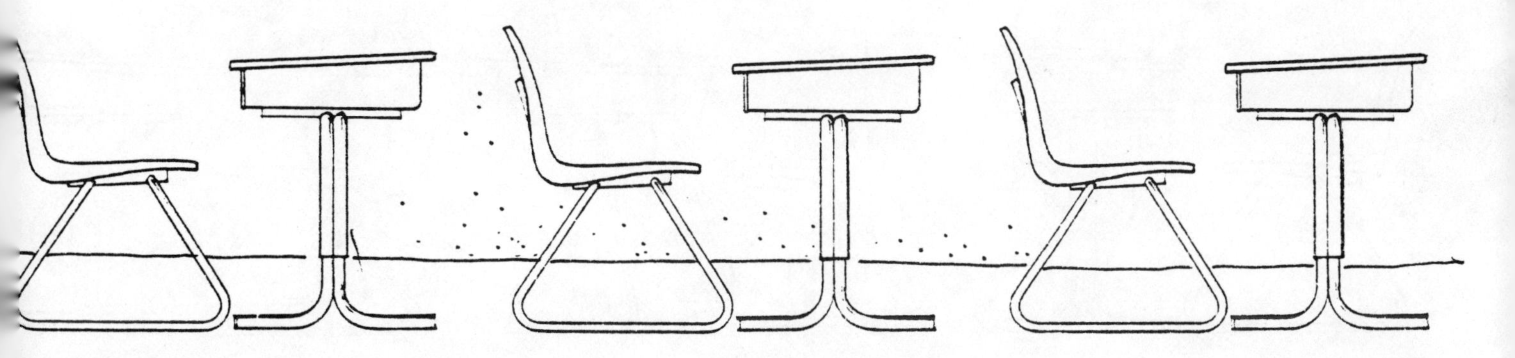

The solar collector you've just read about is part of an "active" solar system because it is made up of moving parts, such as fans and pumps, which need electricity to power them.

And we're not telling how the electricity is made, are we? (Ain't I just awful?)

There is another solar energy system that is growing very popular with people who are interested in using the sun's energy. It is called a "passive" solar system because it has no moving parts.

Don't you consider the billions of racing electronic particles "moving parts"?

In a passive solar system, the sun heats part of the house or "something" that holds the heat, then releases it when it is needed. That "something" can be an inside wall made of cement, brick or stone, called a Trombe wall. Or the "something" could be barrels of water, or a bin filled with rocks.

You know how you enjoy running barefoot in the summer and how, if you run out into the street, it can burn the bottoms of your feet – it's so hot.

Passive solar energy burns your feet?

Even at
night, long after the sun has gone down, the street is still warm to your feet.

Well, that's how passive solar energy works.

I knew it.
It's a
foot-burner.

The street catches the sun, holds it all day and even through part of the night.

Some new houses are being built with twice as many windows facing south as towards the north. The reason?

The sun shines from the south and this gives a house the chance to "collect" the warmth, the same way a solar collector collects sunlight.

Some people are building greenhouses on the south sides of their homes.

That way, the greenhouse collects the sun all day, storing its warmth in water containers or in rocks. This very heat warms the house when it gets cold.

Another kind of passive system is to have water-filled, tall plastic columns behind the windows of a house to absorb the sun's heat.

Also, instead of having a stone or cement wall standing upright, a concrete floor (which is like a wall, only it's lying down)...

can serve as a collector
of sun scooped up by south facing windows.

Sounds like quite a house, with walls lying on the floor and plastic water tanks all over the place.
FREEDOM

It's really simple to make the sun work for you. All you have to do is open the room to sunshine through windows.

Then, when the sun goes down, close off the windows with heavy shades, drapes or indoor shutters. That way you trap the sun (and heat) in the room.

With both the sun
and its heat in the room
the place should be
quite toasty.

Some good points about "active" solar heat are:

(1) It is easier to control the amount of heat you get and where it goes;

(2) It can be added to houses that are already built much more easily than "passive";

(3) There are many solar companies that understand and can install active systems.

Some good points about "passive" solar heat are:

(1) It uses no electricity;

(2) It often uses parts of the house (such as walls or the floor) to store warmth;

(3) It doesn't need as much care as an active system.

INSULATION

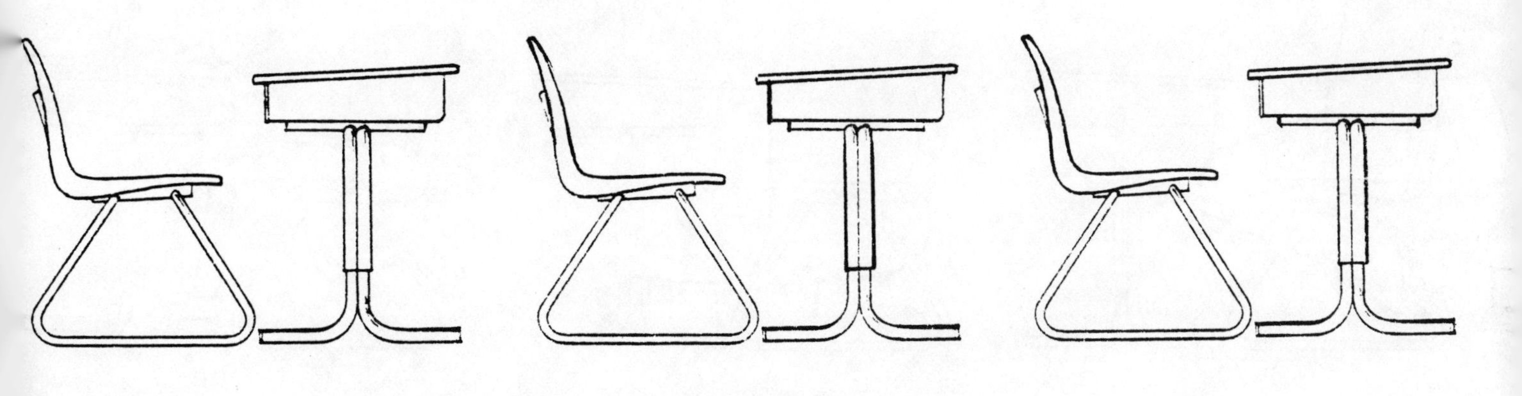

Neither of the two systems, active or passive, will do a house much good if the house is not insulated.

Do you mean a little bit insulated or extra super well-done insulated with every crack sealed against air infiltration?

Is that a new word for you?

It means putting a
blanket between the house and the outside, keeping the heat on one side, the cold on the other. There are many ways of accomplishing that.

And we are about to hear of them, I suspect.
TOY OTA

Perhaps you've seen television commercials about insulating the attic, where a man is seen laying down material on the floor of the attic.

That is an excellent
way to keep the house's heat in the house, where it belongs, instead of in the attic, where nobody lives.

Insulation is any material that traps air and does not permit heat to pass through it easily. It could be cork, paper, different types of plastics that have bubbles caught in them, sponge, rock-wool or glass-wool.

Air itself is insulation, probably the best, and we know that wool (wool winter coat, wool sweater) is an insulator because it does not allow your body heat to escape.

There are many places in a home that need insulation.

It might be a good idea if you went around your own house and checked off the following:

Wait a minute. I'm only a kid.
Why should I do it?
No house needs as much insulation as this school.

(1) Do you feel air leaking in around windows and doors?

(2) Does snow on your roof melt more quickly than that on the roofs of neighboring houses? (That's because the heat in your house is escaping through the ceiling and the roof, and melting the snow.)

(3) Is the hot water heater in your basement covered with insulation?

If it isn't, some of the heat escapes into the basement, where it is wasted.

(4) Are drapes closed and shades pulled when the sun goes down?

That keeps the heat in, the cold out.

(5) Are there storm doors and windows in your house?

Discuss these things with your parents. They will listen closely to what you say because it means saving dollars as well as energy.

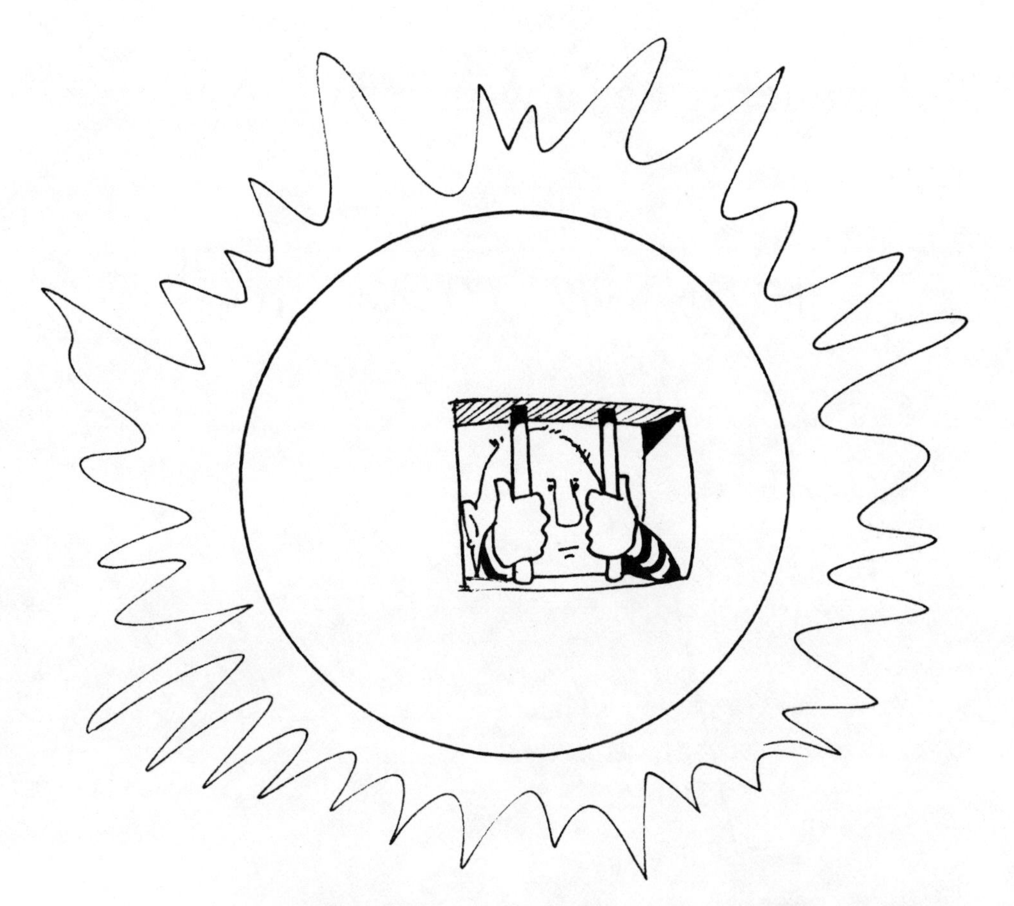

SOLAR CELLS

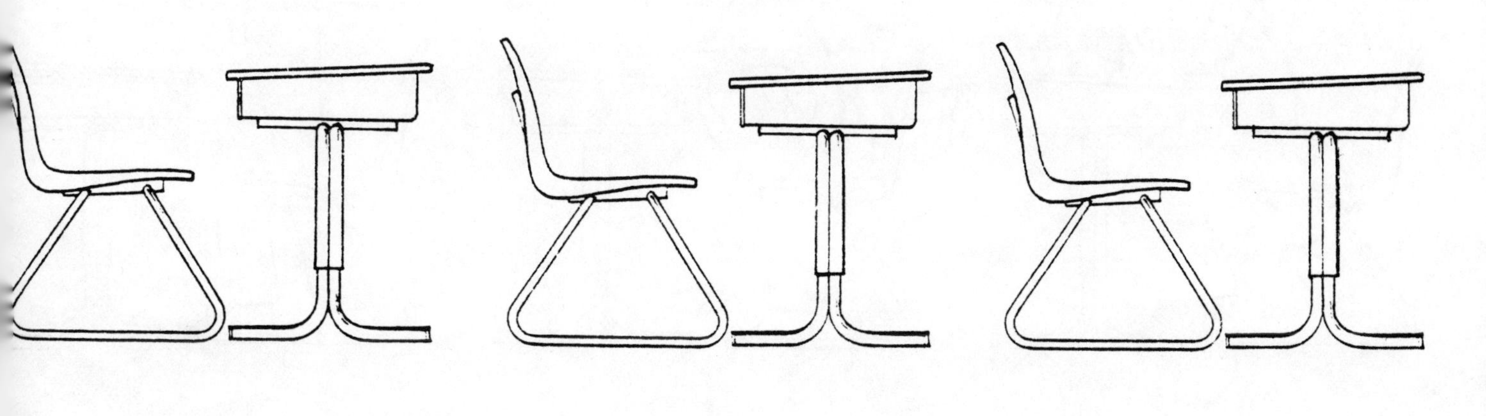

Solar energy has a wonderful future!

One of its most unusual tricks is something called a solar cell. Usually the size of a silver dollar, a solar cell can be as small as the head of a pin.

You may have seen photographs of solar cells and not even have known you were looking at them. Remember seeing a photograph of a space satellite? It had large, wing-like panels coming out of it, didn't it?

You may
have thought those panels helped the satellite to "fly."

Well, in a way, you were right.

Thanks for listening to me, Mrs. Robinson. Most teachers aren't that responsive.

But it's not the way
you think. Those "wings" hold solar cells in them and when light strikes those cells, the special material they are made of changes the light to electricity, needed to power all the electrical equipment on the satellite.

The special material is called photovoltaic *(foe-toe-vol-TAY-ik)* . . .

. . . material, and that's where the big news of solar energy will be coming from in YOUR lifetime! Photovoltaic solar cells can convert light into electricity.

Like light bulbs in reverse.

Imagine this. . . the roof of your house covered with tiny solar cells. That would mean that ALL the electricity you might need (stored in batteries) would come right from your roof.

You could cut the wires
that hook you up to the nearest electric company; you wouldn't need their power anymore because your house would produce its own.

You wouldn't need their power any more because you'd be dead from trying to cut those wires.

Now you know that solar cells already are in use in satellites' "wings." So why can't they be produced now, for use by everyone?

To cook with, for light, to work the vacuum cleaner, to warm every room in the house?

With oil, coal and electricity costing so much, wouldn't it be great if everyone could use solar cells to solve a good part of the energy problem?

Sorry. Not yet.

Solar cells are presently undergoing experiments in business and government laboratories.

The biggest
problem with them now is that they can only convert about 10 per cent of light into electricity.

Until they
can be more efficient than that . . . their price is OUT OF SIGHT.

Now don't get excited,
Mrs. Robinson.
Remember your
blood pressure.

Someday many solar cells will be produced, which will help make them cheaper to buy. They will work much better than they do now, and you will be able to use them to make your own electricity.

Isn't that exciting?

Actually, all three types of solar power systems. . . active, passive and solar cells. . . will be combined to make houses, schools, hospitals, museums, concert halls and businesses warm and comfortable.

It will be done at small cost, thanks to the sun.

It will be done with little pollution, thanks to the sun.

And one of your biggest problems. . . energy. . . will be partly solved, thanks to the sun.

"Thanks to the son.
Thanks to the son.
Thanks to the son."
We daughters never get any credit, do we?

Wait a minute. What's this, the last class? Well, Mrs. Robinson, you really outdid yourself, and I congratulate you. But, before you go, how about showing me a few experiments I can do - things to keep me off the streets and out of trouble?

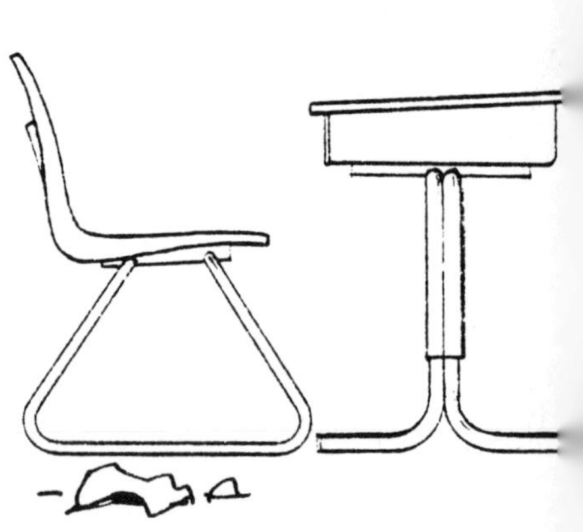

ENERGY EXPERIMENTS

Thanks.

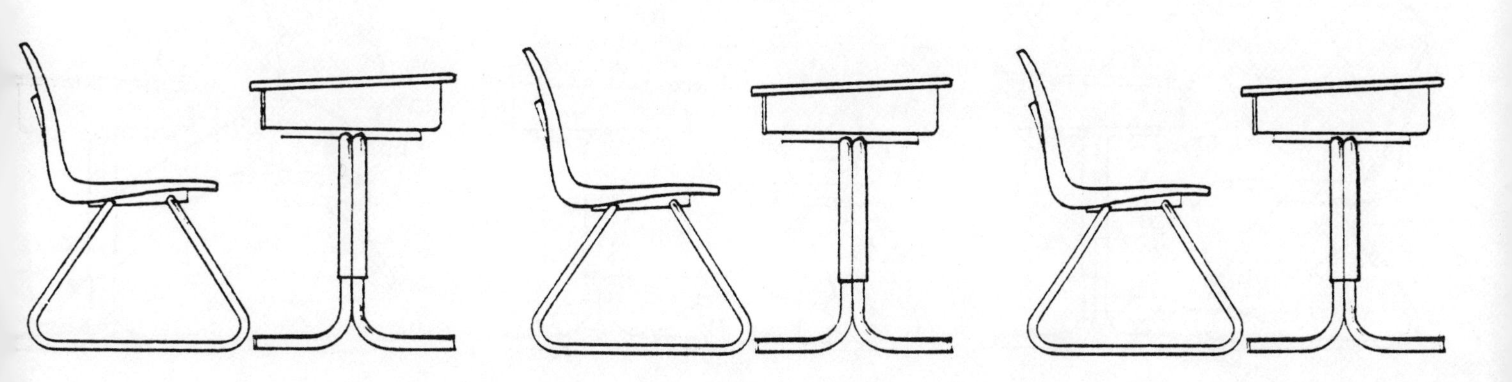

1. Demonstrate Burning Gases:

Light a candle, let it burn a few moments, then blow it out. Quickly bring a lighted match close to the wick, but do not touch it. The flame will appear to "jump" to the wick from your lighted match. The flame is not really jumping, it is igniting the *hot gases* that are still rising from the wick.

2. Demonstrate Pollution from Combustion:
Hold a lighted candle under a clean piece of glass. The spot which forms on the underside of the glass can then be wiped with a tissue.

The dirt (carbon) on the tissue can be circulated in class to show pollution given off by the flame.

The thumb and forefinger can be circulated to show the principles of meat roasting.

3. Demonstrate the Effect of No Sun:
Use two small, potted plants. Put one in a small, light-tight box (a shoe box is fine) and the other beside it, open to the sun on a window sill. Water them equally, as necessary. Note the progressive deterioration of the sunless plant over a period of weeks.

4. Demonstrate Heating of Absorbed Sunlight Vs. Reflected Sunlight:

Paint a small piece of wood (a ruler is fine) with flat black paint. Wrap about half of it with aluminum foil. Place it in the sun for an hour or so. Have students feel the temperature difference between the *absorbing* surface and the *reflecting* surface.

5. Locating South:
Using a simple compass, point various students in different directions and ask each one to locate *north* and *south.*

Mark their findings on the floor with chalk, and see who comes closest to your pre-determined (but hidden) mark.

Now, how about some passive solar experiments? I want to knock a wall flat or put an oven on the roof, something like that. Got any experiments along those lines, Mrs. Robinson, or must I now be satisfied with having learned everything there is to know about solar energy? You're pointing to the door. Is that a message? OK OK I get it. Goodbye. See you later, Mrs. R.

GLOSSARY

ACTIVE
SOLAR SYSTEM

The sun's rays heat the rooftop collector panel through which air is moving. Air at top of the panel is hottest. Air is drawn down via ducts through blower into insulated rock bin which absorbs much of the air's heat, allowing cooler air to rise to collector for another load of energy. When heat from rock bin is needed, simple changes in ducts allow it to enter the rooms.

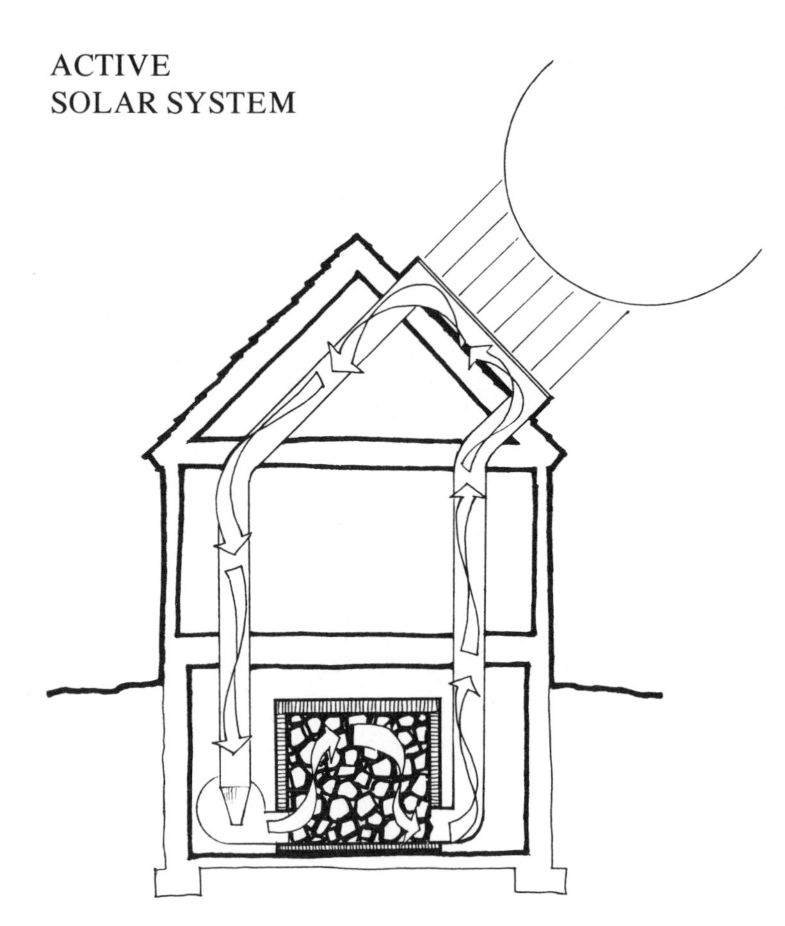

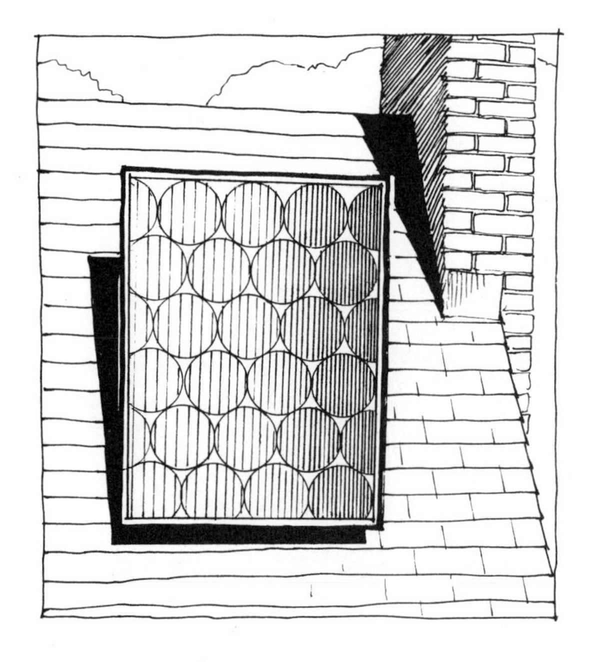

SOLAR CELLS (PV'S)

Photovoltaic panel on roof converts sunlight directly to electricity. Power is either used, stored in batteries, or sold to the power company.

PASSIVE SOLAR SYSTEM

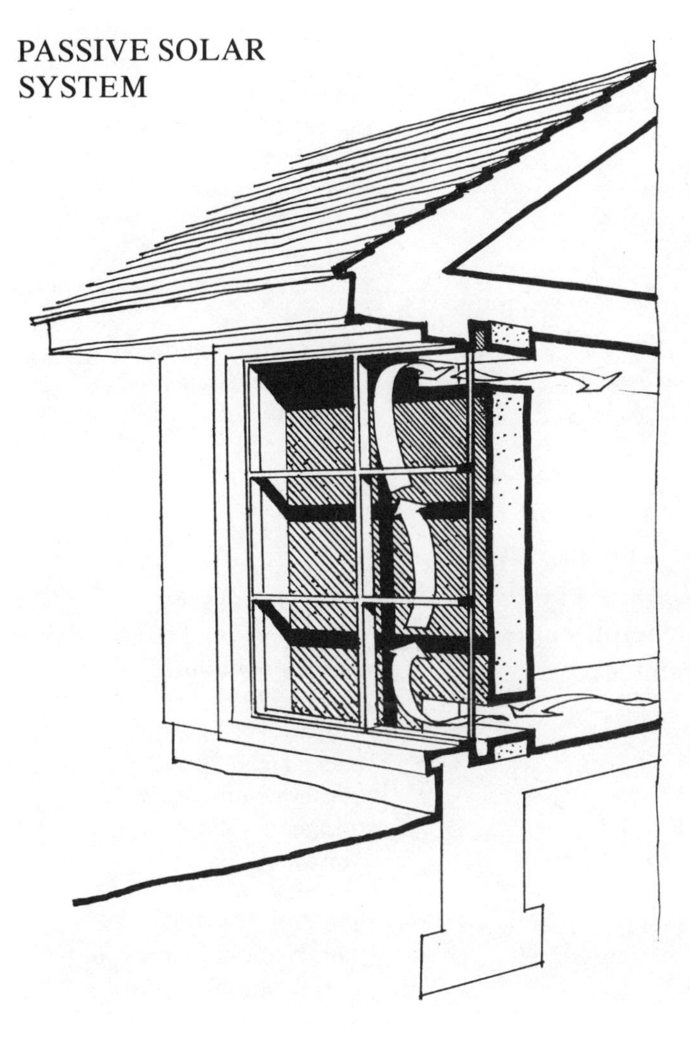

Shaded area shows Trombe wall behind insulating glass. A wall of concrete, brick, stone, or even water drums, painted black, absorbs solar heat on sunny days, and slowly radiates heat to rooms hours later. Top and bottom vents can help circulation of solar-heated air. Flaps on vents prevent unwanted reverse airflow on cold nights.

SOLAR COLLECTOR

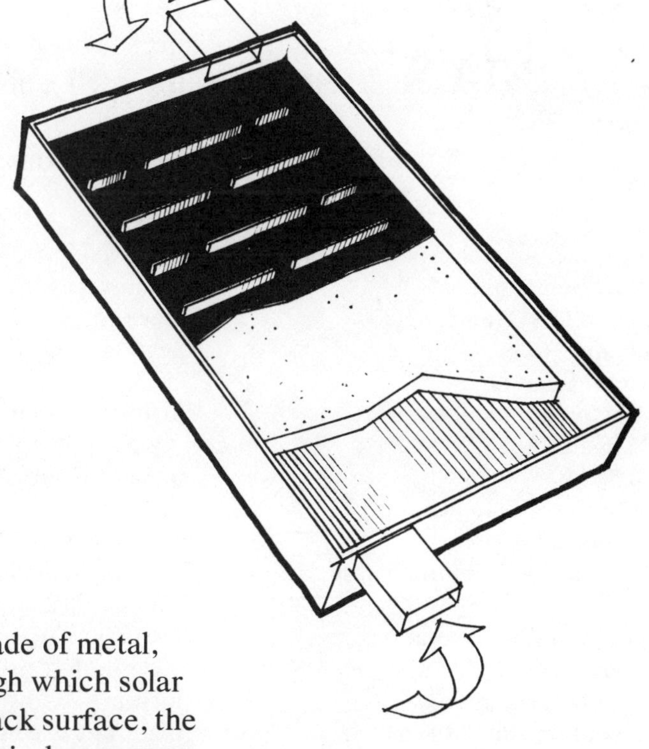

A rooftop collector, usually made of metal, has a glass or plastic cover through which solar energy passes. On striking the black surface, the energy is converted to heat which is drawn away by the moving air. Note arrows showing how air enters and leaves, baffles to improve the transfer of heat to the air, and the insulation behind the black collector surface. Many collectors use water instead of air to carry away the heat, but the general principles are the same.

A solar-powered printing press was a bright idea, but in the 1880's not many people saw the light when it came to popular acceptance.

INDEX